Marc Poling

Walking in the Land of Cars

Marc Poling

Walking in the Land of Cars

Automobile-Pedestrian Accidents in Hillsborough County, Florida

LAP LAMBERT Academic Publishing

Impressum / Imprint
Bibliografische Information der Deutschen Nationalbibliothek: Die Deutsche Nationalbibliothek verzeichnet diese Publikation in der Deutschen Nationalbibliografie; detaillierte bibliografische Daten sind im Internet über http://dnb.d-nb.de abrufbar.
Alle in diesem Buch genannten Marken und Produktnamen unterliegen warenzeichen-, marken- oder patentrechtlichem Schutz bzw. sind Warenzeichen oder eingetragene Warenzeichen der jeweiligen Inhaber. Die Wiedergabe von Marken, Produktnamen, Gebrauchsnamen, Handelsnamen, Warenbezeichnungen u.s.w. in diesem Werk berechtigt auch ohne besondere Kennzeichnung nicht zu der Annahme, dass solche Namen im Sinne der Warenzeichen- und Markenschutzgesetzgebung als frei zu betrachten wären und daher von jedermann benutzt werden dürften.

Bibliographic information published by the Deutsche Nationalbibliothek: The Deutsche Nationalbibliothek lists this publication in the Deutsche Nationalbibliografie; detailed bibliographic data are available in the Internet at http://dnb.d-nb.de.
Any brand names and product names mentioned in this book are subject to trademark, brand or patent protection and are trademarks or registered trademarks of their respective holders. The use of brand names, product names, common names, trade names, product descriptions etc. even without a particular marking in this work is in no way to be construed to mean that such names may be regarded as unrestricted in respect of trademark and brand protection legislation and could thus be used by anyone.

Coverbild / Cover image: www.ingimage.com

Verlag / Publisher:
LAP LAMBERT Academic Publishing
ist ein Imprint der / is a trademark of
OmniScriptum GmbH & Co. KG
Heinrich-Böcking-Str. 6-8, 66121 Saarbrücken, Deutschland / Germany
Email: info@lap-publishing.com

Herstellung: siehe letzte Seite /
Printed at: see last page
ISBN: 978-3-659-63435-2

Copyright © 2014 OmniScriptum GmbH & Co. KG
Alle Rechte vorbehalten. / All rights reserved. Saarbrücken 2014

TABLE OF CONTENTS

LIST OF TABLES

iii

LIST OF FIGURES

CHAPTER ONE:

INTRODUCTION

In discussions of transportation in the U.S., the economic and environmental costs of high dependence on private automobiles and the lack of investment and utilization of public transport often hold center stage. In the process, the most organic mode of human mobility is almost never mentioned. Thus, pedestrianism, and its role in everyday movement, is rarely considered as a mode of "public" transport that needs to be encouraged. In fact, American transport planners and decision-makers, until recently, have largely neglected pedestrians in their master plans, often assuming widespread automobile ownership as the normal mode of urban transportation (Geurs and van Wee, 2004). In contrast, proponents of "new urbanism" and "smart growth" are beginning to emphasize the importance of walkable cities, including pedestrian pathways and multi-use neighborhoods, for tackling traffic congestion and air pollution as well as for enabling a more public notion of the city (Goodwill, 2002).

Pedestrianism, however, brings with it another set of dangers related to safety. In the case of Florida, where dependence on automobiles is very high and public transport systems are often underutilized, the importance of addressing the needs of pedestrians arises from the large number of automobile-pedestrian accidents which characterize its urban areas. In 2011, the Tampa Bay region was ranked as the second most dangerous metropolitan area for pedestrians, behind nearby Orlando (Transportation for America,

1

2011). Tampa Bay has an average annual pedestrian mortality rate of approximately 3.52 per 100,000, even as only 1.7% of workers walk to work. One main reason for this is that Tampa Bay is dominated by lower density and automobile-oriented development patterns, which include high-speed urban arteries that are particularly hazardous for walking (Ernst and Shoup, 2009).

This book considers the extent to which geographic distribution of accidents relates to social inequalities that characterize the Tampa Bay region. Given that access to transport options as well as traffic and roadway characteristics of neighborhoods are likely to vary with differences in income and race, pedestrians face different kinds and degrees of traffic-related dangers in the context of the neighborhood in which they reside. Thus, people with the inability to afford an automobile, commonly residing in poorer urban communities, are the ones that rely on public transport the most, even as these communities have low pedestrian accessibility (Goodwill, 2002). The link between neighborhood characteristics and pedestrian-automobile accidents therefore needs to be examined to gain a better understanding of the environments within which pedestrians dwell in the Tampa Bay region.

Research Question

This research examines the relationship between the occurrence and rate of automobile-pedestrian accidents and various roadway and socioeconomic characteristics of census tracts in Hillsborough county, Florida. This county was chosen as case study since it encompasses the city of Tampa. Drawing mainly on data from the Florida Department of Transportation and the U.S. Census, this study utilized GIS and statistical techniques to (i) map the spatial distribution of automobile-pedestrian accidents and (ii) analyze them through multivariate regression models.

The book is divided into five chapters. The next chapter provides the background to the study situating this research within the larger framework of transport geography, reviewing results from previous studies on pedestrian-automobile accidents and outlining urban geographies of Tampa. Chapter 3 provides the research design, discussing data sources, variables used in the analysis and the steps followed to analyze data. Chapter 4 provides two statistical models which relate accidents to neighborhoods and discusses the findings of the regression analysis. Overall, this book seeks to understand how social inequalities are reflected in the spatial distribution of accidents in Hillsborough County and the implications of this for both accident analysis and urban planning.

Significance of Research

This research delves deeper into pedestrian environments of Hillsborough County and becomes significant in two ways. First, it seeks to draw greater attention to the neighborhoods with high accident occurrence and densities and the links between these neighborhoods and specific income and racial groups. Second, it contributes to the study of how traffic-related dangers affect the mobile experience of urban pedestrians. More broadly, this research can be linked to studies which focus on how unequal distributions of wealth and an urban transport environment geared towards the interests of private vehicle owners ultimately impedes the ability of cities to encourage pedestrianism as a viable option for short-distance mobility.

CHAPTER 2:

LITERATURE REVIEW

Accident analysis has been the subject of much quantitative analysis but these studies have often focused on the accident event itself, including the characteristics of drivers and pedestrians and the roadway features of the accident site. This book seeks to extend such studies by locating accidents in the neighborhoods in which they occur. In this chapter the broader context of transport geography studies is introduced before delving into the accident analyses that were utilized to guide choice of variables for this study. The chapter also delves into the urban geography of Tampa in order to understand how the growth of transport infrastructure and transit planning in the region is linked to its pedestrian environment.

Transport Geography

Transport geography studies the provision of transport systems, the use of those systems for the movement of people and goods, and the relationships between transport and other geographical phenomena (Hay, 2000, p. 855). The sub-discipline fully developed in the 1950s when studies of individual transport modes, such as ports, airports, and railways, were initiated (Hay, 2000, p. 855). Research in transport geography can be classified into 1) network studies, which attempt to describe the "geographical pattern of transport networks," such as roads, railways, canals, etc.; 2) studies of transport nodes and terminals; 3) studies of the "provision of scheduled

4

services" by train, bus, and air; 4) studies of the movement of commodities; and 5) studies of the "movement of people" (Hay, 2000, p. 855). Transport geography also plays a vital role as an "agent or facilitator of geographic change" and seeks to contribute to sustainability in the transport sector (Hay, 2000, p. 856).

A major debate in transport geography swirls around its methodological dependence on quantitative analysis. Beginning in the 1960s, in part motivated by the positivist movement in geography, transport geography experienced a quantitative boost (Hay, 2000, p. 855). The explosion of GIS and its application to transport planning has carved out an empirical niche that is growing in popularity, especially on the government level. This has been useful in gaining more complex understandings of the geometries of transport systems and the socioeconomic identities and everyday experiences of transportation.

Geurs and van Wee (2004) list four major approaches to transport accessibility, which can be defined as the ability to reach transport services and opportunities and is separate from access which is a measure of transport coverage as such (Oregon DOT, 2007): infrastructure-based (efficiency of transport operations), location-based (focused on distance-based measures), utility-based (economic costs and benefits) and person-based. The authors point out that person-based accessibility measures are rarely used in large scale transport initiatives with transport planners and geographers usually preferring the first three measures of accessibility. In turn, two of the other three accessibility perspectives, infrastructure-based and location-based, are favored, with the former utilized frequently by transport planners and the latter often implemented by transport geographers. They explain this situation as reflecting a desire for 'accessibility measures

5

which are easy to interpret for researchers and policy makers (Geurs and van Wee, 2004, p.127). Their study therefore suggests a need to bring in person-based perspectives in ways that can be fitted into the quantitative orientation of existing plans.

Pedestrianism and Transport Geography

The role of pedestrians in transport landscapes is a topic that requires more study. According to Geurs and van Wee (2004), individual pedestrians are not accounted for when transport decisions are made by policy makers, planners, and designers; in their words, pedestrians are "left out of the equation." This exclusion can partly be explained by the fact that pedestrian factors, such as personal choices, preferences, and experiences, require too much time and cost in terms of data collection. However, this could also be a consequence of a need to design traffic systems to accommodate the smooth flow of cars within which pedestrians are mainly viewed as obstacles to the maintenance of steady speeds.

There is a need moreover to connect such studies to the broader political economy of transport-related decision making, including how investment in transport systems reflects class-based priorities and the role played by energy and automobiles multinationals in shaping specific transport systems. More fine-grained qualitative studies are also required in terms of the experiential dimensions of transport landscapes which may shape individual choices regarding use of mass or private transport options and the willingness to walk.

According to recent case studies in environmental justice and public transit, ethnic and racial minorities and persons living in low income households tend to be concentrated in city centers, away from jobs, goods, and services (National Center for

6

Transit Research, 2005). African Americans, other people of color, and low income persons walk, bike, and use transit more than the general population, and are more likely to be victims of death by automobile (National Center for Transit Research, 2005, p. 4). Goodwill (2002) sums it up: "the lack of transportation choice is truly a problem for lower-income persons." Lower income people have the most to gain from transit-oriented development, especially if more effective transport service is the result (Goodwill, 2002, p. 13).

The increased use of automobiles and the decreased accessibility of pedestrians will lead to the continual degradation of poor urban areas, where car ownership is often not a convenient option. Urban locations accessible by automobile will prosper, whereas locations served by public transport will decline (Mackett & Edwards, 1998, p. 231). This process accentuates an ever increasing dichotomy between the more affluent urban residents and the poorer urban communities. This type of segregation can be found in Atlanta, where the public transport system is viewed as a way to increase the total transport capacity into the center of the city (Mackett & Edwards, 1998, p. 234). The urban areas outside the city's center corridor are then ignored regarding accessibility.

At the heart of this disregard may be the way everyday mobility is viewed by planners, developers, decision makers, and society. Walking, the most fundamental method of mobility is taken for granted and underappreciated. Pooley (2009) states that one of the reasons why pedestrians are rarely catered to in transport planning is because 'walking is now viewed principally as a leisure activity rather than transport.' Neglected is the fact that most everyday travel remains short distance, predictable, and adaptable; thus making it possible to reduce car trips in urban areas (Pooley, 2009, p. 149).

Pedestrians, in a world of advanced technology, have come to be known as 'vulnerable road users' or VRUs (Zegeer & Bushell, 2012, p. 3). In the United States, pedestrian fatalities constitute approximately thirteen percent of all traffic deaths (Zegeer & Bushell, 2012, p. 4). That figure cannot be ignored. As more lanes are added to already swollen highways, pedestrians are put at higher risk (Zegeer & Bushell, 2012, p. 5). About seventy percent of pedestrians are killed or injured on road sections, where only about thirty percent are killed or injured at junctions, also known as intersections (Gitelman et al., 2012, p. 65). This demonstrates that pedestrian crossing signals need to be located in more places in addition to intersections. A vehicle has more limitations as to where it can travel and what paths it can take, but a pedestrian has much more freedom of movement. Pedestrian planning should account for this difference in flexible movement. There is no substantive quantitative proof showing that giving pedestrians the right of way decreases accidents (Gitelman et al., 2012, p. 68). As some researchers argue, such as Gitelman et al. (2012), the presence of pedestrians and "their rights" should be strengthened by means of measures reducing travel speeds, enhancing the crosswalk's prominence, or giving separate time for pedestrians to cross (Gitelman et al., 2012, p. 70).

These studies show that a combination of socioeconomic characteristics and roadway features become important in understand the dangers faced by pedestrians. The next section considers studies focused on automobile-pedestrian accidents to further consider the variables that are important to explaining accidents in urban environments.

Automobile-Pedestrian Accident Analysis

Zegeer and Bushell (2012) outline five categories of variables that contribute to explaining pedestrian-related crashes: driver, roadway, pedestrian, vehicle, and

8

demographic factors. This book seeks to focus mainly on socioeconomic and roadway characteristics to explain accidents thus moving away from a focus on the accident event to the accident neighborhood. This section outlines some pertinent studies in the building of a study focused on the neighborhood-level characteristics of accidents.

Census Tract as Unit of Analysis

An important decision in neighborhood analysis of automobile-pedestrian accident analyses is the unit of analysis used to approximate the neighborhood. Using census tracts is a common practice, with some limitations. Census tracts are small enough to reduce the possibility of aggregation errors and ecological fallacies but large enough to potentially avoid impacts that spill over beyond smaller spatial units such as block or block groups (Cottrill & Thakuriah, 2010, p. 1720). According to Ukkusuri et al. (2012), the census tract level provides richer resolution, more data variability, and therefore greater exploratory power for variations in accident frequencies, as compared to zip codes. The census tract level also provides fairly consistent results across the various models that they utilized. There is some support therefore for considering the census tract as a proxy for neighborhood in this research.

Roadway and Traffic Variables

A combination of roadway and traffic variables is often utilized to understand the characteristics of accident sites. In a study of San Francisco, California, Wier et al. (2009) sought to explain automobile-pedestrian accidents using land use, street characteristics, pedestrian exposure proxies, and demographic characteristics within census tracts. Ordinary least squares regression (OLS) was used to model the natural log of the number of automobile-pedestrian injury collisions over a 5-year period. One collision was added

9

to the census tracts with zero collisions reported so they would not be removed from the analysis (Wier et al., 2009, p. 141). The study provided evidence that traffic volume is a major factor in automobile-pedestrian injury collision. In addition, employee and resident populations, arterial streets without public transit, proportions of land area zoned for neighborhood commercial use and residential-neighborhood commercial use, land area, proportion of people living in poverty, and proportion of people aged sixty-five and over are statistically significant predictors of automobile-pedestrian injury collisions. An area-level approach such as this can support pedestrian injury prevention by justifying area-level interventions in development and planning processes (Wier et al., 2009, p. 144).

Some automobile-pedestrian accident analyses investigate collision data at a larger scale, such as intersections in the case of Miranda-Moreno et al. (2011). Their particular study included all injured pedestrians for whom an ambulance was sent on the island of Montreal over a five-year period. Built environment variables in the vicinity of each intersection were assessed using land use, demographic, transit, and road network data. Buffers of various sizes were utilized to find how an intersection's immediate surroundings affected pedestrian activity. Census data at the census tract level and a generated dummy intersection classification variable were also used in the study (Miranda-Moreno et al, 2011, p. 1628). Their results showed that population density, commercial land use, number of jobs, number of schools, presence of metro stations, number of bus stops, percentage of major arterials, and average street length have a powerful association with pedestrian activity. The influence of the built environment on pedestrian collisions at intersections appears to be largely mediated through pedestrian activity and traffic volume. More pedestrian activity and traffic generate more accidents, with traffic volume being the primary cause of collision frequency at intersections (Miranda-Moreno et al, 2011, p. 1633).

Given the frequent lack of traffic data and the difficulties in measuring pedestrian activity and operating speed at the area-wide level, research such as that carried out by Ukkusuri et al. (2012) aimed to estimate the indirect effects of area-wide built environment factors on the distribution of total automobile-pedestrian collisions and fatal collisions at the zip code or census tract level. For purposes of the study, built environment was represented by land use characteristics (employment density, commercial density, area covered by parks, number of schools, etc.), demographics (population density, children population, seniors population, etc.), transit supply (km of bus lanes, number of transit stops, the presence of metro stations, etc.), and road network characteristics (km of streets and major roads, number of intersections, speed limits, etc.) (Ukkusuri et al., 2012, p. 1142). The two dependent variables in the study were severe crashes and fatal crashes. At the census tract level, total pedestrian crashes was also considered as a dependent variable (Ukkusuri et al., 2012, p. 1144). The time period of their study was 2002 to 2006. Their results clearly showed the link of built environment, transit, and road geometric design characteristics on the total and fatal automobile-pedestrian collisions. The proportion of multilane roads is positively associated to both total and fatal pedestrian crashes, highlighting 'the importance of safety interventions that look for a reallocation of road space that can help improving non-motorized transportation safety and modal diversity (Ukkusuri et al., 2012, p. 1150).

The aforementioned studies show that a variety of variables become useful in explaining accidents; the complexity of the built environment and its traffic and roadway

11

features have to be considered. A multivariate regression analysis becomes important here to building explanations that consider this complexity.

Socioeconomic and Demographic Variables

A social and environmental justice approach to transport analysis is another valuable aspect of understanding neighborhood-level characteristics (e.g. Martens et al., 2012; Chakraborty, 2009) Martens et al. (2012) emphasize the importance of including car ownership disparity between low-income and high-income communities. Thus, they argue that transport planners need to focus not only on transport congestion but on issues of access, so that neighborhoods which fall below minimum thresholds in terms of access can receive more attention (Martens et. al, 2012, p. 693).

Fleury et al. (2010) found that people living in underprivileged neighborhoods are at higher risk of being involved in an automobile-pedestrian accident (Fleury et al., 2010, p. 1658). Most of the factors evoked to explain the excess risk of underprivileged residents could be viewed as consequences of economic or social deprivation. Road safety measures in at-risk neighborhoods may act as a form of remediation, but cannot alleviate in its entirety the mobility problems facing lower-income communities. Their results clearly show that socio-spatial differences have a major effect on accident risk. Fleury et al. study therefore calls for a blending of social and road safety policy to combat accidents in high risk neighborhoods (Fleury et. al, 2010, p. 1659).

The social inequalities of urban landscapes are also the outcome of historical processes of settlement and patterns of investment. To situate this study, the next section outlines processes of urbanization and transport development in Tampa and Hillsborough County.

12

Tampa's Urban and Transport Geographies

The advent of the railroad in the U.S. introduced the country to travel it had never before seen. People could settle in cities, such as Chicago, on a whim and physically removed from coastal developments. Goods could be shipped across regions and cities could thrive farther from other production bases. In the cities, new forms of urban transport began to develop in order to ease movement through the city. One such form was the electric streetcar, or trolley, which appeared in cities such as Philadelphia, Los Angeles, and even Tampa. The advent of the electric streetcar sparked the sprawling of urban development in the United States (Gonzales, 2005, p. 346). In the U.S., the motivation for trolleys was backed by a strong desire for land value instead of profit from urban transport development (Gonzales, 2005, p. 347). It comes as no surprise, then, that the first U.S. planners, in essence, were land developers (Gonzales, 2005, p. 348).

Tampa has been a prime example of a developer-influenced planning environment, where gentrification has been glaringly obvious post-WWII. Tampa's entrepreneurs, or 'growth elite,' headed by 'good-old-boy' politics, furthered the trend toward the serial reproduction of post-industrial urban environments (Archer, 1996, p. 250). Place commodification has made social progress difficult in Tampa, as there has been an attempt to 'imagineer' the city's heritage (Archer, 1996, p. 258). Archer (1996) suggests that an inclusionary local politics must be enacted so that local development is not left solely in the hands of public-private 'experts' whose decisions are entirely based on bottom-line, market criteria. Private costs and benefits and the attraction of developers lies in the strong grasp of highway development and maintenance, at the expense of progressive public policies and the socially disadvantaged (Archer, 1996, p. 245).

13

Privatization, combustion engines, and highway development literally derailed

Tampa's once extensive streetcar network. The streetcar system used to be comprised of

190 streetcars covering 53 miles along 11 routes. The streetcar framework was

dismantled in the mid-twentieth century, after the Second World War, and replaced with

highways, freeways, and major automobile arterial thoroughfares (Tampa Preservation,

Inc., 2011). The uprooting of the Tampa streetcar system and the planting of highways

and freeways, such as I-75, I-275, and I-4 have bisected, divided, and displaced several

neighborhoods of Tampa, especially historic African American communities like "The

Scrub," which was dismantled for a freeway. Many of the quarter mile zones surrounding

the Tampa freeways have become wastelands; left vacant, unoccupied, and derelict.

In addition to the negative social and spatial effects of interstate development in

Tampa are the negative environmental effects, as Chakraborty (2009) discusses. The

Tampa Bay Metropolitan Statistical Area (MSA) is plagued by uneven, and unequal,

distribution of health risks from vehicular toxic emissions, based on race and ethnicity.

Most likely to reside in areas of highest risk are households without vehicles, within

census tracts characterized by higher minority populations and lower home ownership.

These census tracts are located near roadways that produce the highest levels of daily

traffic volume and concomitant air toxics (Chakraborty, 2009, p. 693).

High automobile travel and mobility has contributed to even more extreme levels

of sprawl, particularly in newer cities such as Atlanta and Tampa. The U.S. has money

earmarked for roads, has low taxes on energy, and planning is administered on a more

local level (Gonzales, 2005, p. 350). As a result, in 1990, only 74kg of the 4,683kg of

carbon dioxide emissions per capita were from diesel and electrically powered forms of

American transportation (Gonzales, 2005, p. 351). America is by far the leading emitter, per capita and total, of carbon dioxide in the world. American automobile manufacturers have gotten to a level of production and distribution efficiency where they no longer have to increase capacity to serve increased demand (Gonzales, 2005, p. 354). It is no longer costly to place millions of cars, or carbon dioxide emitters, on America's roads. Additionally, in 2004, the U.S. withdrew from the Kyoto Protocol, which is an internationally represented organization that works toward the decrease of global carbon dioxide emissions (Gonzales, 2005, p. 354).

The United Nations (UN) and the World Health Organization (WHO) have both declared the enormous social and economic burden imposed on society by injuries due to road collisions as a major global problem, including pedestrian fatalities and injuries (Feng Wei & Lovegrove, 2012, p. 140). Feng Wei and Lovegrove (2012) posit that different transportation development philosophies significantly influence road safety levels. They also suggest that mode choice is not necessarily determined by demographics or wealth levels, as in the case of Amsterdam, where high income levels do not 'drive' high auto mode use (Feng Wei & Lovegrove, 2012, p. 143). Essentially, the world's transportation problems are not going to be solved by merely fixing symptoms based on specific explanatory factor groups, but by readjusting the global philosophies on transit.

Goodwill (2002) argues that transit oriented design (TOD) needs to be re-instituted in order to make under-served areas more transit friendly. Transit friendly design involves the provision of bus stop amenities for pedestrians and includes safe and convenient pedestrian access to the street and curb cuts. The main purpose of transit-

15

oriented development (TOD) is to enhance mobility by decreasing reliance on the automobile and encouraging use of alternate modes of transportation, such as transit, walking, and biking (Goodwill, 2002, p. 7). Considering street design and proximity of land use, as well as land arrangement, are the keystones of transit orientation.

Research conducted by the Center for Urban Transportation Research (CUTR) suggests that we should carefully consider where public transport options are located in the urban environment. First and foremost, transit nodes, or stations, need to be placed in urban areas that have low household automobile percentages. These areas usually consist of low to moderate income families and immigrant populations. In addition, transit managers should plan responsibly by enhancing transit services within new immigrant populations, who would likely depend largely on public transport for travel (Center for Urban Transportation Research, 1996, p. 23).

Newer cities in the U.S. that have not placed an emphasis on public transport development, like Tampa, are found to have a different public transportation culture as compared to older, more established cities. It is observed in these newer cities that even within three miles of the city center, it is only the low to moderate income populations that utilize public transport. The middle to upper income classes in these areas still primarily drive automobiles to reach their destination(s) (Glaeser et al., 2008, p. 19). This shows that there has been a fundamental shift in the ways public transport planning has developed and how mass transit is perceived by the public. Older cities, such as New York and Washington, D.C., were planned with an emphasis on density, walkability, and public transit. Cities like Tampa and Los Angeles have centered their planning on the automobile, a planning methodology that could be damaging in the long term. There are several structural, social, environmental, and economic tensions that can arise from automobile centricity. The Tampa and Los Angeles downtowns have grown into "nine to five" destinations lacking an established cultural identity or pedestrian presence.

Another newly developed city that acts as a useful urban transport planning case study, Atlanta, has realized that it has a car use and congestion dilemma. Not only is the city attempting to bring more people en masse to the city center, Atlanta has built a metro system to deflate some of the traffic congestion (Mackett & Edwards, 1998, p. 237). The implementation of a metro system in Atlanta has increased public transport ridership, something Hillsborough County, FL has failed to implement.

The focus on private automobiles as the primary mode of transportation thus is part of the historical development of transport systems in Tampa and changes in transit philosophy do not seem likely to emerge in the near future. Accidents linked to pedestrians who are seeking to access public transport or substitute for lack of private vehicles is thus also likely to continue to be a major problem. The rest of this book outlines an analysis designed to understand how automobile-pedestrian accidents are distributed within various neighborhoods in Hillsborough county of which Tampa is a part.

CHAPTER 3:

RESEARCH DESIGN

This chapter provides an overview of the specific steps that were followed in answering the proposed research questions. The contribution of this research emerges from its combination of quantitative analysis and cartographic visualization methods for understanding automobile-pedestrian accidents in Hillsborough County, FL. The chapter begins with a brief introduction to the study area and data sources, followed by a more detailed outline of research methods, including the use of GIS.

Study Area

The case study for this research is Hillsborough County, Florida, which includes the city of Tampa. With a population of approximately 350,000, Tampa is a mid-sized American city. The Tampa Bay metropolitan area consists of roughly 2.75 million people (U.S. Census Bureau, 2010). Most Tampa residents own vehicles and use them as their primary means of transport, inside and outside of city limits. Ninety five percent of daily commuters to Tampa commute via automobile, and only five percent commute via public transport or other means (City-Data, 2011). Tampa has a pedestrian fatality rate twice that of the national average for cities of its size. Every week, pedestrians or bicyclists are injured or killed. The only alternative to walking or bicycling is the bus system. The public transit authority, HART, has recently won performance awards, but it remains an under-utilized and underfunded agency. The Tampa Bay Area Regional Transportation

18

Agency (TBARTA), though, plans to change that with an additional sales tax.

Data Sources

All data collected in this study is publicly available data, used with permission if necessary for the purposes of educational research. The Florida Department of Transportation (FDOT), via the Center for Urban Transportation Research (CUTR), has provided automobile-pedestrian accident data from 2005-2007. The data set contains accidents on roads within FDOT's jurisdiction, also known as state roads. Additional data was obtained directly from the Florida Department of Transportation (FDOT) online data warehouse. Data such as roadway and traffic information was downloaded and imported into ArcGIS for manipulation and presentation.

Demographic data for this study was collected at the census tract level from the 2000 U.S. Census digital data tables. Census data from 2000 was used instead of 2010 census data because, at the time of the accidents investigated in this analysis, the 2000 Census data was the current census information. Moreover, the mismatch between the time period of the accident data and the Census time period would persist irrespective of whether 2000 or 2010 Census data was used. American Community Survey (ACS) data was not used because of the margins of error associated with it.

Public transport nodes, or bus stops since the only mass transit system in Hillsborough County is a bus system, within one quarter mile of each state road were plotted. This transit information was obtained from the Hillsborough Area Regional Transit Authority (HART) via Google's General Transit Feed Specification (GTFS) <http://www.gohart.org/routes/hart/pdf/system_map_local_april_2012.pdf>. Transit nodes were imported, geocoded, and mapped alongside the Hillsborough County state

19

roads, automobile-pedestrian accidents, and census tract layer using ArcGIS overlay and buffering techniques.

Methodology

The analysis of accident data was undertaken in two steps. The first preliminary step was to map the location of various automobile-pedestrian crashes on state roads and gain some familiarity with their spatial distribution. The second step was to build two regression models around two different dependent variables, accident occurrence and accident rate, to explain the relationship between accidents and neighborhood characteristics. These steps are detailed below.

Visualizing Distribution of Accidents

Figure 1 below depicts the census tracts of Hillsborough County, the Florida Department of Transportation (FDOT) state roads within the county, and the locations of automobile-pedestrian accidents on Hillsborough county state roads from 2005-2007. Notice how most of the accidents occurred within the city limits of Tampa.

The greatest numbers of automobile-pedestrian accidents occur in the census tracts of the University Area of Tampa. This is the area approximately 6 to 8 miles north of downtown Tampa, west of and neighbor to the University of South Florida Tampa campus. Between 2005 and 2007, there were a combined 32 state road automobile-pedestrian accidents in the three University Area census tracts that experienced the most incidents. The University Area consists of multi-lane, high volume, high speed thoroughfares with sparse distribution of intersections and traffic signals. The intersections in this area are wide due to the large number of lanes, making the crossing of state road intersections and highways a daunting task for pedestrians.

20

Overall, the census tracts of the North Tampa and East Tampa areas displayed the highest number of automobile-pedestrian accidents. Development, especially transportation infrastructure, is lacking for pedestrians in these areas. Large numbers of lanes and limited places to cross causes pedestrians to be severely disadvantaged compared to automobiles on state roads in these particular areas.

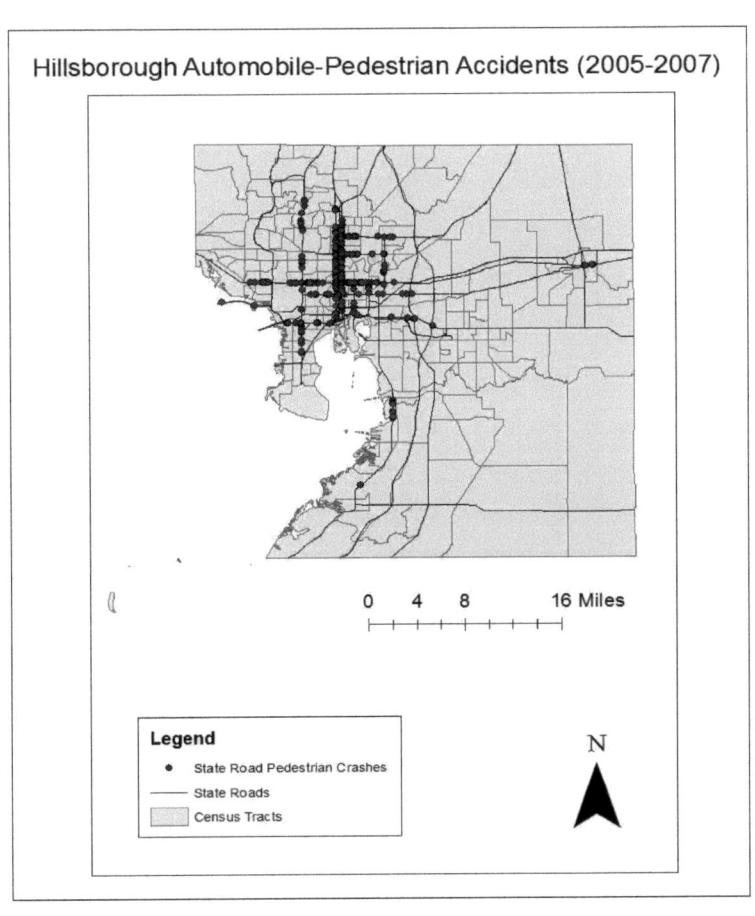

Figure 1: Hillsborough County Automobile-Pedestrian Accidents

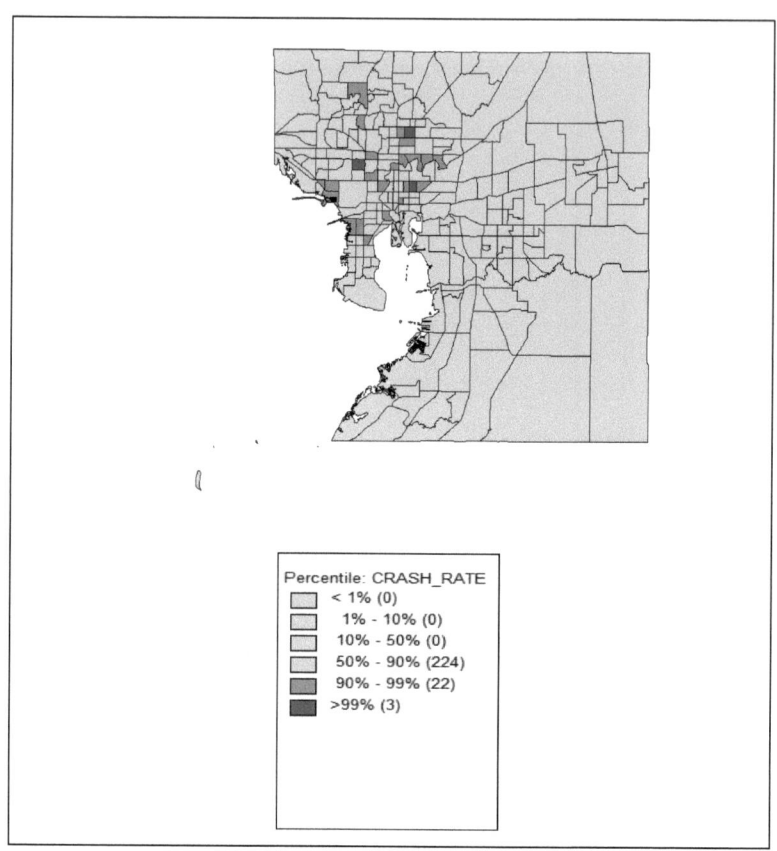

Figure 2: Hillsborough County Automobile-Pedestrian Accident Percentile Map

Figure 2 displays the automobile-pedestrian accident percentile map for

Hillsborough County state roads for the years 2005 to 2007, using GeoDa software. From

the map it can be discerned which census tracts within the county experienced the highest

percentage of automobile-pedestrian accidents. Notice that the highest percentage of

accidents occurred in the northwest portion of the county, where population density is the

greatest.

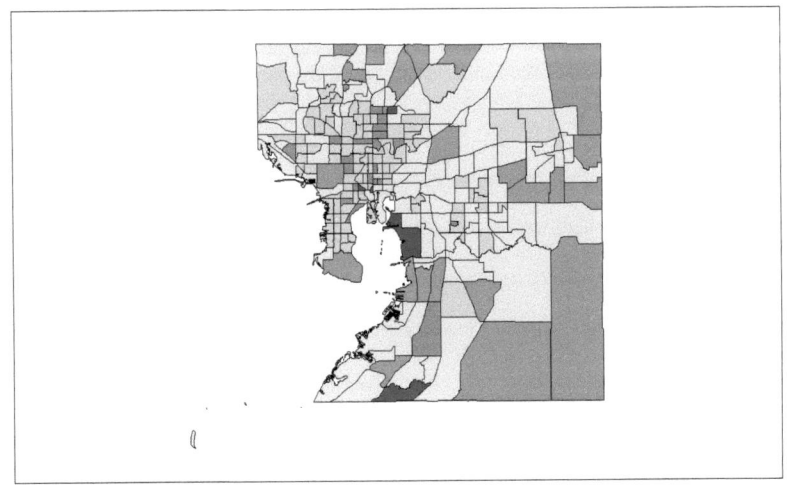

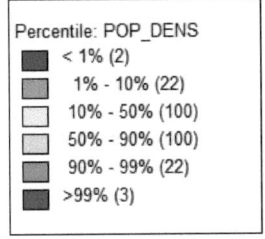

Percentile: POP_DENS
- ◼ < 1% (2)
- ◼ 1% - 10% (22)
- ☐ 10% - 50% (100)
- ◻ 50% - 90% (100)
- ◼ 90% - 99% (22)
- ◼ >99% (3)

Figure 3: Hillsborough County Population Density Percentile Map

The automobile-pedestrian crash rate was calculated by dividing the number of automobile-pedestrian accidents by the number of state road miles; hence, crash rate is defined as the number of automobile-pedestrian accidents per mile of state road. Dark red census tracts in Figure 2 above are tracts that experienced an automobile-pedestrian accident rate within the 99 percentile, or highest crash rates throughout the county. Note that three census tracts fell into this 99 percentile group. Population density is higher in the census tracts with the most accidents compared to the county average of 2,948 persons per square mile.

24

Modeling Accident Occurrence and Rate

To determine the factors influencing the occurrence and rate of automobile-pedestrian accidents on state roads in Hillsborough County, FL from 2005 to 2007, multivariate logistic regression (preceded by stepwise regression) and multivariate linear regression methods were used. First, the relationship between occurrence of accidents (measured as presence and absence of accidents) and the explanatory variables was evaluated using multivariate regression. Then, a linear regression was performed to identify the variables that were most influential in explaining rate of accidents. There were three different combinations of independent factors (traffic components, roadway infrastructure, and demographic and socioeconomic attributes). All statistical analyses used IBM SPSS statistical software (version 20).

Dependent Variables

The dependent variable in the logistic regression analysis was the occurrence of an automobile-pedestrian accident on a state road in each Hillsborough County census tract from 2005 to 2007. If a Hillsborough County census tract experienced an automobile-pedestrian accident on a state road between 2005 and 2007, the census tract received a value of '1' for the dependent value; otherwise it received a '0.' Each census tract within Hillsborough County received a value of '1' or '0,' for a total of 249 values, but only 172 of those tracts were considered in the analysis because not every census tract contained a state road.

A buffer was not placed around each automobile-pedestrian accident event because state roads are quite often neighborhood dividers, especially interstates and freeways. The economic, social, and structural makeup on one side of a state road can be

25

substantially different on the other side of the state road. It was the intention of this study to attempt to isolate certain portions of Hillsborough County, based on several factors (especially socio-economic attributes), that have a high propensity for automobile-pedestrian incidents. Adding a buffer around accidents would pull in accidents to certain Hillsborough County census tracts that may share a state road boundary with another census tract, but have significantly different characteristics from the adjacent tract.

The goal of the dependent variable selection was to help identify factors involved in the presence of automobile-pedestrian accidents and to understand which census tracts within Hillsborough County are potentially more susceptible to automobile-pedestrian accidents. Variables were chosen based on previous studies, with the understanding and acceptance that some variables are highly correlated with one another.

The census tract level was utilized in all of the regressions because the census tract level is a common census area division. Appropriate socioeconomic, demographic, and infrastructural data could be investigated, aggregated, and analyzed at the census tract level. Additionally, census tracts were a unit of area determined to be fitting since pedestrian transportation initiatives and implementations often span beyond the block or block group census units. Consider the Tampa Bay Area Regional Transportation Authority's (TBARTA) 2050 future regional transportation plan, for example. Their plan is to devise vital transportation infrastructure not only spanning census tracts, but counties as well (TBARTA, 2011, p. 10). Analysis at the census block group or census block level would not support cohesion between areas evaluated within a regional transportation master plan.

Most of the census tracts in Hillsborough County span multiple square miles, allowing for significant transportation system analysis to be performed. Any census unit larger than the census tract level would have been too large for the scope and scale of the investigation. The scope of this regression analysis fell between a micro-sized block level and a macro-sized state level. Census tract divisions fit within the county level framework and an acceptable number of census tracts, or number of samples, resulted.

Accident occurrence was calculated by first obtaining the accident data for 2005-2007 from the Florida Department of Transportation. The data provided included all accidents involving vehicles for the state, including automobile-pedestrian accidents. The first task, then, after obtaining the collective data was to eliminate all of the data that was attached to vehicle accidents that did not occur in Hillsborough County and did not involve a pedestrian. In the data set provided, there were data attributes provided, labeled 'COUNTY' and 'F_PED' that signified the county name and the involvement of a pedestrian in an accident, respectively. A selection via attributes was used to only pick the accidents that had a value of "HILLSBOUROUGH" for the 'COUNTY' field. A selection based on attributes was then used to see if a pedestrian was involved. The 'F_PED' field was given a value of '1' if there was a pedestrian involved in the accident, otherwise '0.' In ArcGIS, by selecting accidents based on attributes, only the accidents with a pedestrian value of '1' were chosen for the logistic regression. As a result of the selections, the accident data that remained contained accidents occurring in Hillsborough County that involved a pedestrian.

After the correct accidents were selected, an intersection process was performed in ArcGIS to intersect the Hillsborough County automobile-pedestrian accidents and the

27

census tracts of the county. Using the 'Summarize' functionality in ArcGIS, the 'F-PED' field was summed for each Hillsborough County census tract, based on the 'TRACT_ID' field, or the field that contained the particular tract number of the accident. After the summation procedure, it could then be determined how many automobile-pedestrian accidents occurred in each Hillsborough County census tract from 2005-2007. Discussed earlier, buffer zones were not placed around automobile-pedestrian accidents because of community characteristics that can quickly change from one side of a state road to another or from census tract to census tract. Not every census tract experienced an automobile-pedestrian accident during that time frame, an effect that was expected since some of the Hillsborough County census tracts do not even contain state roads, and as such, did not experience an automobile-pedestrian accident on a state road.

After eliminating Hillsborough County census tracts that did not contain state roads and any outlying automobile-pedestrian accident records, records that displayed a standardized residual value, or 'ZRE' value, greater than the absolute value of 3.0, the final number of samples, or 'N' value, was equal to 169. A 'SR_ACCIDENT_LOGISTIC' value of '1' was provided for each Hillsborough County census tract with an automobile-pedestrian accident between 2005 and 2007, a value of '0' was given to any other sample record.

The data collection, selection, aggregation, and manipulation outlined were the basis and origin of the dependent variable in the logistic regression analysis, also known as 'SR_ACCIDENT_LOGISTIC.' In all, 76 census tracts out of the 249 within Hillsborough County experienced at least one automobile-pedestrian accident on a state road from 2005 to 2007.

During the logistic regression, a residual analysis was performed to determine any skewness or heteroskedasticity inherent in the variable relationships. The results of the residual analysis can be viewed in the appendix of this report. Table 2 below outlines the number of automobile-pedestrian accidents in each Hillsborough County census tract. The census tracts are ordered by 'TRACT_ID,' which was a field provided by and defined by the U.S. Census Bureau. Table 3 explains the summary statistics of the dependent and independent variables, and also notes the Pearson's r value for each variable used in the logistic regression, all of them significant at the p<.01 significance level (two-tail). Other summary statistics included in the table are variable minimum, maximum, mean, and standard deviation.

Table 1. Summary Statistics for Dependent and Independent Variables – Logistic Regression

Variables	Min	Max	Mean	SD	Pearson's r
Dependent:					
State Road Accident Occurrence	0	1	0.3	0.458	
Traffic Components:					
Avg. State Road Speed Limit (mph)	30	70	49.7	8.2	-0.564**
Annual Avg. Daily State Road Traffic	2163	126900	43318	20995	0.206**
Roadway Infrastructure:					
State Road Bus Stop Dens. (per mi^2)	0	91.85	9.25	16.8	0.639**
Total State Road Length (miles)	0	92.4	11.217	14.6	0.208**
Demographic Attributes:					
Avg. Travel Time to Work (min.)	11.8	39.6	25.77	4.432	-0.270**
% Families below FPL	0	73.8	10.42	10.22	0.429**

**p<.01; *p<.05 (two-tail)

The dependent variable used in the linear regression of the Hillsborough County census tracts with automobile-pedestrian accidents was state road accident rate. This variable was equal to the number of automobile-pedestrian accidents per square mile of state road. Accidents per square mile of state road was used instead of accidents per square mile of area because this study focused on state roads and the accidents

29

surrounding them, not on all of the roads within Hillsborough County. If this study incorporated the accidents on all roads within the county, the number of accidents per square mile of area would have been used as the dependent variable. Using accidents per square mile of area would have been misleading. State road accident rate was also chosen because after determining which census tracts contained accidents and using logistic regression to identify census tracts of vulnerability, evaluating which census tracts experience "clustering" accident effects was the next logical step.

There were a total of 74 census tracts, or samples, used in the linear regression of census tracts with accidents. Although 76 census tracts experienced accidents, two of those census tracts lacked complete data values for the purposes of the regression. In all, nine explanatory variables were included in the regression, which are discussed in further detail later in this chapter. More than nine variables were first considered for inclusion, but after preliminary regression results, some of the variables up for consideration were considered to be not influential based on the standardized coefficients, t-values, and significance levels.

Note that based upon the residual plots of the proposed dependent variables, it was not determined that any dependent variable transformations would be needed in this analysis. Conical or quadratic patterns were not detected in the residual plots of any of the regressions. Additionally, none of the proposed independent variables required transformations, which are required to address skewness or kurtosis in the frequency distribution, based on their frequency histograms or scatter plots. For these reasons, Table 1 does not list any transformed variables. Interaction terms were not considered.

Table 2. Summary Statistics for Dependent and Independent Variables – Linear Regression of Hillsborough County Census Tracts with Accidents

Variables	Min	Max	Mean	SD	Pearson's r
Dependent:					
State Road Accident Rate	0.26	25	4.635	5.12	
Traffic Components:					
Avg. State Road Speed Limit (mph)	30	57.8571	44.75	5.548	-0.288*
Annual Avg. Daily State Road Traffic	6733	126900	48146	21647	0.052
% Workers Commuting via Vehicle	36.2	88.3	72.46	10.11	-0.167
Roadway Infrastructure:					
State Road Bus Stop Dens. (per mi^2)	0.7	91.85	24.78	22.05	0.503**
Total State Road Length (miles)	0	92.4	15.74	17.29	-0.059
Avg. State Road Lane Count	1	3.5	2.54	0.45	-0.295*
Demographic Attributes:					
Avg. Travel Time to Work (min.)	12.3	39.6	24.21	4.77	0.147
% Families below FPL	0	73.8	16.6	12.9	0.380**
% Non-White Residents	3.36	98.32	39.3	26.36	0.237*

**p<.01; *p<.05 (two-tail)

Explanatory Variables

There were a total of six explanatory, or independent variables used in the logistic regression. The six variables were considered part of three broad statistical categories: traffic components, roadway infrastructure, and demographic attributes. For the purposes of this study, the three categories described provided an appropriate contextual framework for the research revolving around census tract automobile-pedestrian accidents.

Average State Road Speed Limit

Average state road speed limit (mph) was considered as an explanatory variable in the logistic regression because vehicle speed plays a role in not just the severity of an automobile-pedestrian accident, but the odds of such an accident. Obviously, at lower speeds, vehicles have a shorter stopping distance when vehicle brakes are applied. Slower vehicle speeds can make it easier for pedestrians to cross traffic. Traffic controls, including speed limits, are cited as significant factors within pedestrian accident analysis,

31

as noted by Zegeer and Bushell (2012) and Feng Wei and Lovegrove (2012). As such, speed limit was considered for this study.

The average state road speed limit data, categorized under the Traffic Components label for the purposes of this study, was obtained from the Florida Department of Transportation according to the most recent data. The data was accessed from FDOT's public traffic data library. The information came bundled for the entire road system of Florida. This study was only concerned with the traffic components of state roads within Hillsborough County, so the state data had to be filtered to accommodate the study. In order to do this, the speed limit data layer was intersected with a Hillsborough County census tract layer obtained from the Florida Geographic Data Library (FGDL). The speed limit data was further filtered by selecting only the state roads within Hillsborough County. What remained was a layer of data consisting of state road segments within Hillsborough County and their corresponding speed limits. The next step was to average the state road section speed limits over each Hillsborough census tract. The summarize feature of ArcGIS was used to average the speed limit of each state road segment within each census tract.

It should be noted that average roadway speed limit figures do not account for driver behaviors such as speeding and driving while intoxicated or under the influence of a narcotic. The actual average vehicle speeds are probably proportionally higher than the speed limit data suggests.

Average Annual Daily State Road Traffic (AADT)

Average annual daily state road traffic levels, like average state road speed limit, fell into the Traffic Components category. Zegeer and Bushell (2012) regard vehicle

32

volume as an important factor in the analysis of pedestrian crashes. The sheer numbers of vehicles on particular roads can influence the number of automobile-pedestrian crashes. In Hillsborough County, percent of driving commuters can factor largely into vehicle volumes, as the percent of commuters travelling via vehicle is exceptionally high within the county. The lack of public transportation options also contributes to high vehicle volumes in the county. High vehicle volumes can make it difficult for pedestrians to cross roadways, especially when there are multiple lanes to cross. Often, high vehicle volumes drive infrastructural development as well. If there are many cars on the road, the built roadways have to accommodate vehicles, potentially taking away from the attention given to pedestrians and pedestrian infrastructure.

The process for collecting the average annual daily state road traffic levels was very similar to the process used to finalize the average state road speed limit data. The data came from the Florida Department of Transportation (FDOT) roadway information library according to the most recent data. The data had to be compressed in ArcGIS to allow for the analysis of only the state road segments within Hillsborough County. For each specific, predefined segment of state road within Hillsborough County that had a particular vehicle volume count attached to it, the average annual daily traffic numbers were averaged over all of the segments within each census tract. It is important to note that average annual daily state road traffic levels, upon the completion of new roads, are not static statistical numbers.

State Road Bus Stop Density

According to Zegeer and Bushell (2012), speed limits and traffic density, like bus stop density, when evaluating automobile-pedestrian crashes, are categorized under roadway/traffic factors. In this study, however, speed limits and traffic volume were considered within a separate traffic category. Bus stop density in this study, though, did fall under a Roadway Infrastructure label. Several studies (including Lovegrove, 2012, et al.) suggest that the number of automobile-pedestrian accidents does increase with the increase in bus stops. However, the same studies show that overall accident numbers do increase, but the vehicle to pedestrian proportion decreases, causing the overall number of accidents per pedestrian percentage to decrease. Therefore, bus stop density figures as they relate to pedestrian crash analysis can be deceiving. It cannot be flatly concluded that adding more bus stops makes mobility worse for pedestrians. The research is available to support the idea that large bus stop numbers means greater safety for pedestrians as a whole.

The Hillsborough County bus stop data was obtained from the Hillsborough Area Regional Transit Authority (HART), via public Google transit data (GTFS). All of the bus stops for Hillsborough County were collected. The bus stops were then selected based on proximity to Hillsborough County state roads. Only the bus stops within one quarter mile, or roughly the distance under normal circumstances that the average person will walk to public transit, of Hillsborough state roads were selected. The total number of bus stops near state roads in each Hillsborough census tract was calculated using the summarize functionality of ArcGIS. To find the state road bus stop density, for each census tract, the total number of bus stops near state roads was divided by the census tract area.

Total State Road Length

Total state road length is another variable that was included in the Roadway Infrastructure category. A study by Feng Wei and Lovegrove (2012) also included a road length variable in the pedestrian crash analysis. Naturally, just as bus stop increases

usually amount to greater amounts of automobile-pedestrian crashes, the same could

generally be said of increased road lengths. For the purposes of this study, the unit of

measurement for road length was equal to miles.

State road data was provided for the entire state of Florida, so as with some of the

other variables, the data was compressed into a layer that depicted only the state roads

within Hillsborough County. Each state road segment within Hillsborough County had a

shape length associated with it. The length initially provided was in meters so the unit

was converted to miles to remain consistent with the other units of measurement within

the study. A summation of the state road shape lengths for each Hillsborough Census tract

was performed to calculate the total state road length of each tract.

It should be noted that state roadway density, or the length of state road per square

mile of Hillsborough census tract, was not considered in the study to avoid unit

redundancy. Roadway density is also a difficult variable to quantify because in the case of

automobile-pedestrian crashes, roads could be considered as polygon areas as opposed to

line features, their typical designation. Determining the locations of automobile-

pedestrian accidents then becomes difficult because of the complexities of roadway

density calibration.

Average Travel Time to Work

Average travel time to work, as Zegeer and Bushell (2012) agree, is an integral

factor in tracking commuter diagnostics as they relate to pedestrian crashes. The umbrella

group for this variable was Demographic Attributes. In the case of Hillsborough County,

we know that average travel time to work, because of the high vehicle commuter

35

numbers, largely relates to how many minutes it takes, on average, for people to drive to work in the morning. A longer commute time may mean several things, including but not limited to: dense traffic, long road lengths, and low population/employment density. If commuters are experiencing dense traffic and/or long roadway lengths on their way to work, it could mean that pedestrians are experiencing the negative consequences of those occurrences, especially people walking to work or walking to public transit.

The commuter travel time data was acquired from the 2000 U.S. Census. The unit of measure was minutes since average commute times are in the minute range and not the hour range. Previously mentioned, 2000 Census information was used because it provided the most accurate demographic situation at the time of the recorded accidents in this study; the 2010 Census was not in existence during the 2005-2007 timeframe. A future study would incorporate accident data from a later time period, such as 2010 to 2012, with the ability to utilize the 2010 Census data.

The online public census record tools allow for the "drilling down" of information. For example, it was possible to filter in real time the census data tables to match exactly with Hillsborough County, FL. Provided was the commuter information table just for Hillsborough County census tracts. This process allowed for quick manipulation of the census data and easy export, then import, into IBM SPSS Statistics 20. The import was based on a data merge by Hillsborough census tract ID number. With the addition of roads, average travel time to work could increase or decrease over time.

Percent of Families below the Federal Poverty Level (FPL)

The Federal Poverty Level (FPL), or the Federal Poverty Income Guideline (FPIG), is an income threshold developed by the federal government. Any person or family considered below the FPL can be eligible for certain public benefits, such as supplemental income assistance, food assistance, child care assistance, utility assistance, etc. A large proportion of Americans, and Hillsborough County residents, live below the FPL. Percent of families living below the FPL, for this study and the study by Zegeer and Bushell (2012), was grouped under the Demographic Attributes heading.

This variable was an integral factor in the regression analysis because one of the primary goals of the study was to attempt to formulate a relationship between automobile-pedestrian accidents in Hillsborough County and some of the socio-demographic characteristics associated with the area. Since two demographic variables, average travel time to work and percent of families below the FPL, were included in the regression, the regression analysis was more statistically balanced and not based primarily on physical, more tangible variables associated with infrastructure or traffic patterns.

As with the commuter travel time data, the FPL data for each census tract was extracted from the online 2000 U.S. Census data tables. Other income-based data was extracted from the 2000 Census data, but the FPL data eventually was considered for the regression analysis because it had the greatest influence on the regression statistics compared to the other income-related data. The data was supplied in percentage form and was kept in that form as opposed to being converted to decimal form, a form that would have been less transparent in the regression output.

More variables were used in the linear regression because statistical model predictability was not as much of a concern for the linear regression compared to the logistic regression. The three additional variables are defined below. The three additional

37

variables help describe the automobile-pedestrian accidents in relationship to neighborhood level traits, aiding in painting a clearer picture of the reality of transport in Hillsborough County, FL.

Percent of Workers Commuting via Vehicle

Vehicle ownership figures and commuting habits are very common factors considered in automobile-pedestrian accident analysis. The percentage of workers commuting via vehicle was selected because it not only is highly correlated to vehicle ownership, but it also helps describe the commuting environment of an area or region. A large percent of the population commuting via vehicle, as is the case in Tampa and its surrounding county area, suggests that there are less than desirable public transit options available found to be suitable or efficient for commuting to and from work. As more vehicles are introduced to the environment, pedestrian safety becomes more of a concern, especially during the rush hour periods prior to and after regular business hours; early in the morning and early in the evening.

Percentage of workers commuting via vehicle also sheds light on the overall tendency of a population to drive to other destinations, separate from work. As someone drives home from work, they are more likely to drive to a store or for groceries, or some other activity, since they are already in the privacy and isolation of their vehicle. A high percentage of driving commuters exasperates the inflation of a "car culture."

The data for this variable was made available by the U.S. Census Bureau, based on the 2000 Census data. Other commuter-related data considered, but not used in this analysis, included percent of workers commuting via alternate means, such as public transit, walking, or other forms of transport. Percent of workers commuting via vehicle was ultimately utilized because it has been found to add the most to overall pedestrian danger when compared to other commuting methods.

Average State Road Lane Count

One of the most dangerous roadway infrastructure implementations for pedestrians is an abundance of multi-lane highways. As roadway lanes increase, so do vehicle speeds and roadway widths. At slow vehicle speeds, pedestrians are still very vulnerable, so as vehicle speeds increase, pedestrian vulnerability becomes an even greater concern. The concern builds when roadway widths, and therefore intersection widths, increase in size. As lane numbers increase, the pedestrian must be mindful of additional rows of automobiles. It also takes a pedestrian longer to cross a road when there are several lanes. And as several studies have shown, increasing roadway lanes does not correlate to lessened traffic congestion. In fact, multiple studies have shown that adding more lanes to a major road, an interstate for example, causes greater vehicle congestion.

The state road lane count data was obtained from the Florida Department of Transportation's (FDOT) data library. The data provided lane information for all of the Florida state roads. The dataset was condensed so that it only included the lane information for state roads within Hillsborough County, FL. Average lane count numbers tend to increase over time.

Percent Non-White Residents

High volumes of automobile-pedestrian accidents are often highly correlated to areas of predominantly non-white populations. Furthermore, non-white populations are often highly correlated to percentage of residents living below the federal poverty line. As previously mentioned, some of the variables used in the linear regressions had a high correlation value with each other, including percentage of people living below the federal poverty line and percentage of non-white residents. The correlation between the two variables, although high, was ignored for the purposes of the linear regressions because both factors are vitally important for not only understanding the inherent, unfortunate relationship between minorities and poverty levels, but for better understanding how race and income affect many aspects of life, including pedestrian safety.

The variable representing the percentage of non-white residents is directly defined by how census takers answered the race question on the 2000 Census survey. Individuals categorized as "white" on the census survey are not considered as part of the percentage of non-white residents. The complete accuracy of this data is unclear because topics such as race and ethnicity can be sensitive for many people, difficult for some people to describe, or improperly interpreted or extrapolated by the U.S. Census Bureau.

Statistical Analysis

Roadway infrastructure, traffic and demographic and socioeconomic attributes, were merged with the census tract boundaries to form the informational foundation for subsequent regression analysis. The demographic characteristics of the bounding census tracts were related to the accident, transit, and roadway layers through a regression-based model, in an effort to contemplate the significance of the influence that transportation features have on the mobility of underprivileged city dwellers. The appropriate multivariate logistic regression model was fitted to the data after evaluating three separate models.

40

A comparison was made between the accident locations and the socio-demographic characteristics of the surrounding area. Accidents were determined based on the relationships of multiple variables. Hillsborough County, FL consists of 249 census tracts. Only 172 of them were considered in the logistic regression analysis. The census tracts excluded were ones that did not contain any Florida state roads and therefore did not experience any state road automobile-pedestrian accidents during the study period.

A binary logistic regression analysis was performed on the variables presented earlier in the chapter. There were three major steps involved in the logistic analysis. The first step was to determine the correlations between all of the variables and identify multicollinearity concerns. Any correlation with a Pearson's r value of 0.7 or greater or -0.7 and less was considered unacceptable for the purposes of this study. The correlation statistics helped informed the variable inclusion/exclusion process in later analysis steps.

The second step of the logistic analysis was performing a Forward: LR stepwise binary logistic regression on the collected variables. Stepwise regression was utilized to determine the variables that had the greatest statistical impact on the dependent variable. The process involved running stepwise regressions until the results contained approximately ten variables that could be deemed statistically impactful. Since stepwise regression does not account for correlations, some of the combinations of variables returned were considered unacceptable when used together in a standard regression. One variable from each unacceptable pair was removed, with the removal decision based on the variable with the lower significance value and prominence throughout previous studies.

41

The third step in the logistic analysis involved running a binary logistic regression with the set of variables obtained from the stepwise regression. Variables were removed and three models were considered until the final model provided an acceptable fit; meaning the model was statistically significant, predicted a high percentage of accident prone census tracts, contained variables that were all statistically significant at the 0.1 confidence level (or a 90% confidence interval) and were not correlated with one another at a Pearson's r value of 0.7 or greater or -0.7 and lower.

The final part of the logistic analysis, within the third primary step, was evaluating the residual statistics of the regression and eliminating outliers and influential records appropriately. The logistic model was developed with the intention of predicting automobile-pedestrian accident occurrence, so that is why three of the data skewing samples, or outliers, were removed from the analysis. An outlier was considered a record that had a standardized residual value, or ZRE value, greater than the absolute value of 3. It was determined that three of the final 172 samples were unacceptable based on their standardized residual values. Those samples were removed from the analysis altogether. Considering Cook's, Leverage, and dfBeta values, there were not any samples that had influential statistics significantly different than the other samples. The results of the final logistic regression model can be observed in Chapter 5 and the IBM SPSS Statistics 20 output can be viewed in the appendix.

After the logistic regression was performed, a linear regression analysis was performed on the Hillsborough County census tracts that experienced an automobile-pedestrian accident during the study period. Seventy-four census tracts were included in the regression. There were two primary steps involved. The first step was to determine

42

which variables, chosen based on contribution to explanation of process and by existing studies, either had a high correlation to other variables considered or had a lower statistical impact on the overall model fit than other variables. The elimination process involved evaluating the standard coefficients, t-values, and significance values of variables. Some of the variables in the linear regression had a Pearson's r correlation value of greater than 0.7, but those variables were not removed from the regression because they were deemed as relevant and pertinent to understanding how the variables, in relation to one another, described the dependent variable in question. Once the variables for the linear regression had been identified, the second step involved running an ordinary least square (OLS) regression on those variables.

The next chapter details the results of the two regression analyses in terms of the relationships between accident occurrence and rate variables and various neighborhood characteristics.

CHAPTER 4:

RESULTS AND DISCUSSION

The two dependent variables in this study, accident occurrence and accident rate, display significant relationship to roadway, traffic and socioeconomic characteristics. The specific nature of these relationships is described and interpreted in this chapter.

Accident Occurrence

The first regression was a logistic regression that used automobile-pedestrian accident occurrence as the dependent variable. According to the results of the logistic regression, for every bus stop added within one quarter mile of a state road, per square mile, in a Hillsborough County census tract, the odds of an automobile-pedestrian accident occurring in the census tract increases by $(1.107 - 1 = 0.107)$ or 10.7%; when the other independent variables are held constant. This result is significant at the 0.01 level, meaning that adding one bus stop per square mile near a state road will increase the chances of a census tract automobile-pedestrian accident by 10.7%. State road bus stop density is positively correlated to automobile-pedestrian accident likelihood.

According to the logistic regression results, for every 1% increase in families living below the Federal Poverty Level (FPL) in a Hillsborough County census tract, the odds of an automobile-pedestrian accident occurring in the census tract increases by $(1.137 - 1 = 0.137)$ or 13.7 percent; when the other independent variables are held constant. Each 1% increase in families living below the FPL will increase the chances of

44

an automobile-pedestrian accident occurrence. This is an important point that help

explains the social reality of marginalized mobility.

Table 3. Logistic Regression of Automobile-Pedestrian Accidents on State Roads in Hillsborough County, FL from 2005-2007

Variables	Model 1	Model 2	Model 3
Avg. Annual Daily State Road Traffic (AADT)	1.000	1.000	1.000
	(20.375)***	(14.444)***	(15.358)***
Avg. State Road Speed Limit (mph)	0.763	0.780	0.808
	(45.237)***	(21.129)***	(15.153)***
State Road Bus Stop Density (per square mile)		1.118	1.107
		(16.388)***	(11.667)***
Total State Road Length (miles)		1.044	1.035
		(7.138)***	(4.307)**
% Families below Federal Poverty Level			1.137
			(6.986)***
Avg. Travel Time to Work (minutes)			0.877
			(3.278)*
N	169	169	169
X^2	95.319	128.530***	138.617***
-2LL	134.633	101.422	91.335
R^2_{CS}	0.431	0.533	0.560
R^2_N	0.580	0.716	0.753

Note: log-odds coefficients with chi-square values in parentheses; ***$p<.01$; **$p<.05$; *$p<.10$ (two-tail)

Accident Rate

The second part of the overall regression analysis involved the linear regression of

census tracts with pedestrian crashes. The dependent variable for this portion of the study

was accident rate, or the number of accidents per square mile of state road.

According to the results of the linear regression for census tracts with accidents, for every one unit increase in the state road bus stop density of a census tract, the accident rate will increase, keeping all other variables constant. According to the linear regression of all census tracts, for every one unit increase in state road bus stop density of a census tract, the total number of accidents within a tract will increase, holding all other variables constant.

It should be noted that the increase in automobile-pedestrian accident chances and figures with the addition of bus stops does not mean that adding more bus stops makes traveling increasingly dangerous for pedestrians. The net effect of adding bus stops is generally positive for pedestrians because the pedestrian crash to automobile proportion drops, and that is the figure that matters the most in evaluating the level of pedestrian safety in an area.

According to the linear regression results for census tracts with accidents, for every 10 percent increase in the percent of families living below the poverty level in a census tract, the accident rate will increase by 3.16 percent, holding all other variables constant. Based on the linear regression of all census tracts, the percent of families living below the federal poverty line may not impact the total number of accidents in a census tract, maintaining the constant values of all other variables. Important to note is that the variable for percent of families living below the poverty level is potentially not significant in relation to total number of accidents when evaluated with the other eight variables in the regression. This can lead us to believe that factors related to income and poverty levels can help provide an indication of census tract accident vulnerability or proneness, but may not be able to substantively assist in understanding number of accidents. It would be easier to understand the relationship between actual number of accidents and several dependent factors if the dependent variable, number of accidents, was converted to a rate. The actual number of accidents per each census tract is not large, on average, which causes ineffectiveness and inaccuracies in linear regressions using number of accidents as the dependent variable; differentiation becomes unclear.

Table 4. Linear Regression of Automobile-Pedestrian Accidents on State Roads in Hillsborough County, FL from 2005-2007 – Census Tracts with Accidents

Variables	Standardized Coefficient (t-value)
Avg. Annual Daily State Road Traffic (AADT)	0.322 (2.705)***
Avg. State Road Speed Limit (mph)	-0.140 (-0.970)
% Workers Commuting via Vehicle	0.169 (1.193)
State Road Bus Stop Density (per square mile)	0.314 (2.190)**
Total State Road Length (miles)	-0.091 (-0.828)
Avg. State Road Lane Count	-0.368 (-3.301)***
% Families below Federal Poverty Level	0.316 (1.619)*
Avg. Travel Time to Work (minutes)	0.062 (0.557)
% Non-White Residents	-0.063 (-0.358)
N	74
Condition Index	64.163
Adjusted R-squared	0.348
F-statistic	5.336***

Note: State road accident rate as dependent variable; ***p<.01; **p<.05; *p<.10 (two-tail). Condition Index value indicates multi-collinearity (correlation between independent variables). Correlation was anticipated and accepted for the purposes of this study.

Overall Discussion

The results provide further proof that pedestrians in underserved communities are in greater danger of falling victim to automobile-related accidents. Furthermore, neighborhoods predominantly living below the Federal Poverty Line (FPL) deserve a closer look when it comes to pedestrian planning, infrastructure, and safety. The results show that living in impoverished areas begets pedestrian vulnerability.

The logistic model suggests that there is a possibility of correctly predicting if a Hillsborough census tract will be prone to automobile-pedestrian accidents. Linear regression of the census tracts with automobile-pedestrian accidents suggests that four factors are notably influential on accident rate: average annual daily traffic on state roads, state road bus stop density, average state road lane count, and percent of families living below the federal poverty level. The same regression suggests that three factors—total state road length, mean travel time to work, and percent non-white residents—are not notably influential on accident rate.

The diversity of the variables included in the regression analysis demonstrates that predicting the occurrence of and explaining automobile-pedestrian accidents using only neighborhood level characteristics in Hillsborough County census tracts is complicated and multi-faceted. There are many causal factors associated with automobile-pedestrian accidents, and this study shows that a holistic approach to predicting and explaining such accidents is necessary. One category of explanatory variables is insufficient for investigations of this manner. The complexity of the analysis is one reason behind slow implementation changes. Overall, 88.2 % of the values for the logistic dependent value were predicted correctly by the final logistic regression model.

48

This overall accuracy was compared to 84.6% and 81.1 % for Models 2 and 1, respectively.

An intriguing outcome of the logistic regression analysis is that as average state road speed limits and average travel times to work increase, the odds of an automobile-pedestrian accident occurring in a Hillsborough County census tract decrease by 19.2% and 12.3%, respectively. The state road speed limit results could be partially explained by the presence of interstates, roads with high speed limits and low numbers of pedestrians. The work travel time results could be partly explained by driving commuters that are travelling far distances outside of their census tracts of residence to reach their place of employment. These types of commuters, in census tracts of lower employment densities, could be spending more time driving in other census tracts than their own.

The study results also demonstrate that by decreasing the percentage of census tracts families living below the Federal Poverty Level (FPL) and decreasing the total state road length within a Hillsborough County census tract, automobile-pedestrian accidents in the census tract will decrease, holding all other variables constant. As we can see, poorer census tracts with state road arteries running through them are highly susceptible to automobile-pedestrian accidents. Pedestrian-related accidents are not just about infrastructural issues or traffic patterns and flows; they are also related to socioeconomic and demographic features of the respective population. Pedestrian awareness initiatives and infrastructure improvements need to be focused more on underprivileged areas. Not only is obtaining employment, hence obtaining a vehicle, difficult in these areas but the act of travelling on foot from one place to another poses a risk. The oppositions to quality of life mount on top of one another in underserved communities, causing a snowball effect. Dangerous walking conditions are all but one of the many challenges facing neglected neighborhoods.

Average state road lane count, and its relationship to census tract accident rate, needs to be considered based on this study's linear regression of census tracts with automobile-pedestrian accidents. As more lanes are added to roads, or as more roads with several lanes are added to a census tract, accident rate increases. Hillsborough County accidents are shown to cluster around multi-lane roads. One study of interest would be one that attempts to evaluate the suggested number of lanes per road so as to maintain safe walking distances and scenarios for pedestrians. Additionally, a spatial regression study that evaluates the spatial clustering of accidents would add supplemental value to the linear regression of census tracts with accidents used in this study. Not only could a future study investigate spatial clustering, but temporal clustering as well.

A significant result from this entire analysis is the disparity in percent non-white residents in Hillsborough County, FL census tracts with automobile-pedestrian accidents and those without. Tracts with accidents are comprised, on average, with 40% minorities, or non-whites, whereas tracts without accidents are comprised of only 20%. There appears to be a substantial inequality present at the tract level. Often discussed is the economic divide between whites and non-whites in America, but there is a basic mobility divide that requires further attention.

CHAPTER 5:

CONCLUSION

This study sought to understand the pedestrian environment in Hillsborough County and the city of Tampa by analyzing relationships between automobile-pedestrian accidents and neighborhood characteristics. The main findings and limitations of this study are discussed in this concluding chapter.

Accident Occurrence

Automobile-pedestrian accidents are more likely to occur in neighborhoods that showcase statistical variable means that are comparatively high in average annual daily traffic (AADT), state road bus stop density, total state road length, percent of families living below the Federal Poverty Line (FPL), average state road lane count, and percent of non-white residents. Conversely, automobile-pedestrian accidents are more likely to occur in neighborhoods that showcase statistical variable means that are comparatively low in average state road speed limit, mean travel time to work, and percentage of workers commuting via automobile.

Additionally, automobile-pedestrian accidents are more likely to occur in census tracts where there is an increasing trend in the values of average annual daily traffic (AADT), state road bus stop density, and percent of families living below the Federal Poverty Level (FPL). A deceasing trend in the value of state road speed limit also signifies the greater likelihood of automobile-pedestrian accident occurrences in

51

Hillsborough County census tracts.

Based on a combination of the findings from the logistic regression and the means comparison, the primary influential variables affecting the occurrence of automobile-pedestrian accidents in Hillsborough County are average annual daily traffic (AADT), state road bus stop density, and percent of families living below the Federal Poverty Level (FPL). Secondary influential variables are total state road length, average state road lane count, and percent of non-white residents.

Accident Rate

Accident rate is related to neighborhoods that are increasing in average annual daily traffic (AADT), average state road lane count, and state road bus stop density. A secondary influential factor related to accident rate is the percent of families living below the Federal Poverty Level (FPL). There are multiple duplicate influential variables across the means comparison, logistic regression, and linear regression analyses.

Most notably, the factors of average annual daily traffic (AADT) and state road bus stop density are primary influencers across the entire study, regardless of statistical technique and attributing variables. This finding demonstrates the complexities in striking the balance between controlling the daily number of drivers and supporting bus transportation systems as they relate to automobile-pedestrian accident figures. Commonly, as bus stop density increases, daily traffic decreases, but that is not always the case. This study helps prove a valuable point by raising a fundamental question: how do we as a society and community effectively implement bus transportation systems that increase the number of public transit riders, decrease the number of automobiles on the road, and in turn decrease the pedestrian crash to automobile ratio?

Limitations and Future Studies

This study can form the basis for a more expanded study of neighborhoods and accidents in the Tampa Bay region. Given the dangers that lurk for pedestrians on Tampa's roads, it is imperative that such expanded studies be undertaken. When conducted with attentiveness to social inequalities, these studies can then contribute to building a more just urban environment for all Tampa residents.

A comprehensive study that focuses more heavily on the pedestrian factors, or in other words the victims of automobile-pedestrian accidents, would be particularly intriguing. Pedestrians that are struck by vehicles are victims, whether they are at fault or not. By nature, pedestrians are much more physically vulnerable than vehicles and their operators; therefore, pedestrian victims need to be investigated in further detail in order to understand more clearly the problems they face on the road. Pedestrian deaths are a serious matter and the stories behind the people struck by vehicles need to be heard by the general population.

Additionally, a study, like the one performed here, that addresses automobile-bicycle accidents could be helpful for further explaining the transportation environment of Hillsborough County. Many Hillsborough bicyclists suffer the same fate as pedestrians. Due to the inherent differences between the transportation of pedestrians and bicyclists, bicyclists were not addressed in this study, but their safety is just as important as pedestrians.

Smart transportation planning starts with the pedestrian. Walking is the most basic form of travel; it has several health, environmental, and economic benefits. Pedestrians must be acknowledged, catered to, and protected. This study assists in providing evidence that the occurrence of automobile-pedestrian accidents can be predicted to a certain extent. Certainly not every automobile-pedestrian accident can be predicted, but if we can filter out varied, strongly influential factors influencing the occurrences, magnitudes, and numbers of accidents, we could provide a better, safer service to the community. Along with smart growth, proper transportation planning can ease some of the spatial manifestations of social injustice. Spatial justice is a component of social justice; it is time that the dialogues on spatial justice begin to inform discussions on transport planning.

LIST OF REFERENCES

Allen, W.B., Liu, D., & Singer, S. (1993). Accessibility Measures of U.S. Metropolitan Areas. *Transportation Research-B,* Vol. 27B, No. 6, pp. 439-449.

Archer, K. (1996). Packaging the place: Development strategies in Tampa and Orlando, Florida. *Local economic development in Europe and the Americas, 3rd Edition.* Edited by Christophe Demaziere and Patricia A. Wilson. Mansell Publishing Limited, New York, NY, 1996, pp. 239-263.

Chakraborty, J. (2009). Automobiles, Air Toxics, and Adverse Health Risks: Environmental Inequities in Tampa Bay, Florida. *Annals of the Association of American Geographers,* Vol. 99, No. 4, pp. 674-697.

Cho, G., Rodriguez, D. A., & Khattak, A. J. (2009). The role of the built environment in explaining relationships between perceived and actual pedestrian and bicyclist safety. *Accident Analysis and Prevention,* Vol. 41 (2009), pp. 692-702.

City-Data. (2011). Work and Jobs in Tampa, Florida (FL) Detailed Stats: Occupations, Industries, Unemployment, Workers, Commute. Web. 27 Oct. 2011. <http://www.city-data.com/work/work-Tampa-Florida.html#meansOfTransportationToWork.>

Cottrill, C.D. & Thakuriah, P.V. (2010). Evaluating pedestrian crashes in areas with high low-income or minority populations. *Accident Analysis and Prevention,* Vol. 42 (2010), pp. 1718-1728.

Davis, Gary A. (2004). Possible aggregation biases in road safety research and a mechanism approach to accident modeling. *Accident Analysis and Prevention,* Vol. 36 (2004), pp. 1119-1127.

Ernst, M. & Shoup, L. (2009). Dangerous by Design: Solving the Epidemic of Preventable Pedestrian Deaths (And Making Great Neighborhoods). *Transportation for America: Surface Transportation Policy Partnership,* Washington, DC.

Feng Wei, Vicky & Lovegrove, Gord. (2012). Sustainable road safety: A new (?) neighborhood road pattern that saves VRU lives. *Accident Analysis and Prevention,* Vol. 44 (2012), pp. 140-148.

Fleury, D., Peytavin, J.F., Alam, T., & Brenac, T. (2010). Excess accident risk among residents of deprived areas. *Accident Analysis and Prevention,* Vol. 42 (2010), pp. 1653-1660.

Geurs, K.T. & van Wee, B. (2004). Accessibility evaluation of land-use and transport strategies: review and research directions. *Journal of Transport Geography,* Vol. 12, pp. 127-140.

Gitelman, V., Balasha, D., Carmel, R., Hendel, L., & Pesahov, F. (2012). Characterization of pedestrian accidents and an examination of infrastructure measures to improve pedestrian safety in Israel. *Accident Analysis and Prevention,* Vol. 44 (2012), pp. 63-73.

Glaeser, E.L., Kahn, M.E., & Rappaport, J. (2008). Why do the poor live in cities the role of public transportation. *Journal of Urban Economics,* Vol. 63, No. 1, pp. 1-24.

Goetz, A.R., Vowles, T.M., & Tierney, S. (2009). Bridging the Qualitative-Quantitative Divide in Transport Geography. *The Professional Geographer,* Vol. 61, No. 3, pp. 323-335.

Gonzales, G.A. (2005). Urban Sprawl, Global Warming and the Limits of Ecological Modernization. *Environmental Politics,* Vol. 14, No. 3, pp. 344-362.

Goodwill, J. (2002). BUILDING TRANSIT ORIENTED DEVELOPMENT IN ESTABLISHED COMMUNITIES. *Center for Urban Transportation Research,* University of South Florida, Tampa.

Hay, Alan. (2000). Transport Geography. *The Dictionary of Human Geography, 4th Edition.* Edited by R. J. Johnston, Derek Gregory, Geraldine Pratt, and Michael Watts. Blackwell Publishers Inc., Malden, MA, 2000, pp. 855-856.

Hillsborough Area Regional Transit Authority (HART). (2009). Web. 09 Oct. 2010. <http://gohart.org/>.

Iacono, M., Krizek, K.J., & El-Geneidy, A. (2010). Measuring non-motorized accessibility: issues, alternatives, and execution. *Journal of Transport Geography,* Vol. 18, pp. 133-140.

Lee, C. & Adbel-Aty, M. (2005). Comprehensive analysis of vehicle-pedestrian crashes at intersections in Florida. *Accident Analysis and Prevention,* Vol. 37 (2005), pp. 775-786.

Liu, S. & Zhu, Xuan. (2004). Accessibility Analyst: an integrated GIS tool for accessibility analysis in urban transportation planning. *Environment and Planning B: Planning and Design 2004,* Vol. 31, pp. 105–124.

Mackett, R.L., & Edwards, M. (1998). The impact of new urban public transport systems: Will the expectations be met? *Transportation Research Part A: Policy and Practice*, Vol. 32, No. 4, pp. 231-245.

Maggio, E. (2006). Access to public transportation: An exploration of the national household travel survey appended data [electronic resource]. Tampa, FL: University of South Florida.

Martens, K., Golub, A., & Robinson, G. (2012). A justice-theoretic approach to the distribution of transportation benefits: Implications for transportation planning practice in the United States. *Transportation Research Part A*, Vol. 46 (2012), pp. 684-695.

Middleton, Jennie. (2010). Sense and the city: exploring the embodied geographies of urban walking. *Social and Cultural Geography*, Vol. 11, No. 6, pp. 575-596.

Miranda-Moreno, L.F., Morency, P., & El-Geneidy, A.M. (2011). The link between built environment, pedestrian activity and pedestrian-vehicle collision occurrence at signalized intersections. *Accident Analysis and Prevention*, Vol. 43 (2011), pp. 1624-1634.

Murray, A.T., Davis, R., Stimson, R.J., & Ferreira, L. (1998). Public transportation access. *Transportation Research Part D: Transport and Environment*, Vol. 3, No. 5, pp. 319-328.

National Center for Transit Research (U.S.), National Urban Transit Institute (U.S.), & University of South Florida Center for Urban Transportation Research. (1996). *Journal of public transportation* [electronic resource].

National Center for Transit Research. (2005). Case Studies in Environmental Justice and Public Transit Title VI Reporting Final Report TCRP J-06, Task 47 FDOT BD 549-10 (September 2005).

National Center for Transit Research. (2007). Public Transit in America: Analysis of Access Using the 2001 National Household Travel Survey. *Center for Urban Transportation Research* (February 2007).

Oregon Department of Transportation. (2007). *Accessibility and Mobility Differences*. Web. 04 Feb. 2007. <http://www.oregon.gov/ODOT/SUS/accessibility_mobility.shtml>.

O'Sullivan, D., Morrison, A., & Shearer, J. (2000). Using desktop GIS for the investigation of accessibility by public transport: an isochrone approach. In *International Journal of Geographical Information Science*, Vol. 14, No. 1, pp. 85–104.

Pooley, C.G. (2009). Mobility, History of Everyday. *Lancaster University, Lancaster, UK,* c. 2009 Elsevier Ltd. All rights reserved. pp. 144-149.

Prato, C.G., Gitelman, V., and Bekhor, S. (2012). Mapping patterns of pedestrian fatal accidents in Israel. *Accident Analysis and Prevention,* Vol. 44 (2012), pp. 56-62.

Rodrigue, J.P., Comtois, C., & Slack, B. (2006). *The Geography of Transport Systems.* New York, NY: Routledge.

Sanchez, T.W., Shen, Q., & Peng, Z.R. (2004). Transit Mobility, Jobs Access and Low-income Labour Participation in US Metropolitan Areas. *Urban Studies,* Vol. 41, No. 7, pp. 1313-1331.

Schneider, R.J., Ryznar, R.M., & Khattak, A.J. (2004). An accident waiting to happen: a spatial approach to proactive pedestrian planning. *Accident Analysis and Prevention,* Vol. 36 (2004), pp. 193-211.

Tampa Bay Area Regional Transportation Authority (TBARTA). (2011). A Connected Region for Our Future. *Master Plan Vision,* Adopted June 24, 2011, pp. 1-30. Retrieved Oct. 12, 2011 at <http://www.tbarta.com/update>.

Tampa Preservation, Inc. (2011). Tampa's Streetcar System. Web. 13 Oct. 2011. <http://tampapreservation.com/2011/01/tampa-streetcar-system/>.

TampaGov. (2009). Web. 07 Dec. 2010. <http://www.tampagov.net/dept_Land_Development/information_resources/data_ and_statistics/census_2000/census_2000_map.asp>.

Transportation for America. (2011). Dangerous by Design 2011. Web. 27 Oct. 2011. <http://t4america.org/resources/dangerousbydesign2011/states/worst-metros/>.

Ukkusuri, S., Miranda-Moreno, L.F., Ramadurai, G., & Isa-Tavarez, J. (2012). The role of built environment on pedestrian crash frequency. *Safety Science,* Vol. 50 (2012), pp. 1141-1151.

United States Census Bureau. (2010). Population Estimates. Web. 27 Oct. 2011. <http://www.census.gov/popest/cities/.>

Van Lange, P.A.M., Vugt, M.V., Meertens, R.M., & Ruiter, R.A.C. (1998). A social dilemma analysis of commuting preferences: The roles of social value orientation and Trust. *Journal of Applied Social Psychology,* Vol. 28, No. 9, pp. 796-820.

VAN VUGT, M., VAN LANGE, P.A.M., & MEERTENS, R.M. (1996). Commuting by car or public transportation? A social dilemma analysis of travel mode judgments. *European Journal of Social Psychology,* Vol. 26, No. 3, pp. 373-395.

Wier, M., Weintraub, J., Humphreys, E.H., Seto, E., & Bhatia, R. (2009). An area-level model of vehicle-pedestrian injury collisions with implications for land use and transportation planning. *Accident Analysis and Prevention,* Vol. 41 (2009), pp. 137-145.

Zegeer, Charles V. and Bushell, Max. (2012). Pedestrian crash trends and potential countermeasures from around the world. *Accident Analysis and Prevention*, Vol. 44, 2012, pp. 3-11.

Zhao, F., Chow, L.F., Li, M.T., Ubaka, I., & Gan, A. (2003). Forecasting Transit Walk Accessibility: Regression Model Alternative to Buffer Method. *Transportation Research Record: Journal of the Transportation Research Board*, Vol. 1835, pp. 34-41.

APPENDIX A:

ADDITIONAL TABLES

Table A1. Final Logistic Regression Model Variable Statistics

Statistics

		SR accident rate (1 or 0)	average annual daily SR traffic (integer)	average SR speed limit (mph)	percent of families below federal poverty level (integer)	SR bus stop density (per square mile)	total SR length (miles)	mean travel time to work (minutes)
N	Valid	246	170	169	246	246	246	246
	Missing	0	76	77	0	0	0	0
Mean		.30	43318.36584	49.683185	10.4195	9.249347	11.216941	25.7683
Std. Error of Mean		.029	1610.260917	.6289012	.65154	1.0707423	.9292111	.28257
Median		.00	42069.57145	49.000000	7.3500	.901830	5.602044	25.4000
Mode		0	24500.0000^a	45.0000	1.00^a	.0000	.0000	23.80^a
Std. Deviation		.458	20995.23369	8.1757159	10.21894	16.7939361	14.5741063	4.43192
Variance		.210	440799837.6	66.842	104.427	282.036	212.405	19.642
Skewness		.895	.660	.326	2.511	2.481	2.128	.102
Std. Error of Skewness		.155	.186	.187	.155	.155	.155	.155
Kurtosis		-1.208	1.057	-.104	9.732	6.341	6.354	.518
Std. Error of Kurtosis		.309	.370	.371	.309	.309	.309	.309
Range		1	124737.5000	40.0000	73.80	91.8555	92.3875	27.80
Minimum		0	2162.5000	30.0000	.00	.0000	.0000	11.80
Maximum		1	126900.0000	70.0000	73.80	91.8555	92.3875	39.60
Sum		73	7364122.194	8396.4582	2563.20	2275.3393	2759.3674	6339.00

a. Multiple modes exist. The smallest value is shown

Table A2. Final Logistic Regression Model Variable Correlations

Correlations

		SR accident rate (1 or 0)	average annual daily SR traffic (integer)	average SR speed limit (mph)	percent of families below federal poverty level (integer)	SR bus stop density (per square mile)	total SR length (miles)	mean travel time to work (minutes)
SR accident rate (1 or 0)	Pearson Correlation	1	.206**	-.564**	.429**	.639**	.208**	-.270**
	Sig. (2-tailed)		.007	.000	.000	.000	.001	.000
	Sum of Squares and Cross-products	51.337	361540.958	-383.792	492.076	1203.015	339.301	-133.985
	Covariance	.210	2139.296	-2.284	2.008	4.910	1.385	-.547
	N	246	170	169	246	246	246	246
average annual daily SR traffic (integer)	Pearson Correlation	.206**	1	.213**	.002	.026	.055	-.030
	Sig. (2-tailed)	.007		.005	.975	.737	.478	.696
	Sum of Squares and Cross-products	361540.958	74495172556	6121489.122	90335.797	1744305.297	2925057.047	-488932.276
	Covariance	2139.296	440799837.6	36437.435	534.531	10321.333	17308.030	-2893.090
	N	170	170	169	170	170	170	170
average SR speed limit (mph)	Pearson Correlation	-.564**	.213**	1	-.360**	-.539**	.223**	.384**
	Sig. (2-tailed)	.000	.005		.000	.000	.004	.000
	Sum of Squares and Cross-products	-383.792	6121489.122	11229.511	-5101.654	-13918.841	4609.097	2408.220
	Covariance	-2.284	36437.435	66.842	-30.367	-82.850	27.435	14.335
	N	169	169	169	169	169	169	169
percent of families below federal poverty level (integer)	Pearson Correlation	.429**	.002	-.360**	1	.602**	.110	.017
	Sig. (2-tailed)	.000	.975	.000		.000	.084	.794
	Sum of Squares and Cross-products	492.076	90335.797	-5101.654	25584.526	25313.879	4027.669	185.932
	Covariance	2.008	534.531	-30.367	104.427	103.322	16.439	.759
	N	246	170	169	246	246	246	246
SR bus stop density (per square mile)	Pearson Correlation	.639**	.026	-.539**	.602**	1	.097	-.152*
	Sig. (2-tailed)	.000	.737	.000	.000		.128	.017
	Sum of Squares and Cross-products	1203.015	1744305.297	-13918.841	25313.879	69098.891	5840.168	-2779.770
	Covariance	4.910	10321.333	-82.850	103.322	282.036	23.837	-11.346
	N	246	170	169	246	246	246	246
total SR length (miles)	Pearson Correlation	.208**	.055	.223**	.110	.097	1	.036
	Sig. (2-tailed)	.001	.478	.004	.084	.128		.572
	Sum of Squares and Cross-products	339.301	2925057.047	4609.097	4027.669	5840.168	52039.121	573.592
	Covariance	1.385	17308.030	27.435	16.439	23.837	212.405	2.341
	N	246	170	169	246	246	246	246
mean travel time to work (minutes)	Pearson Correlation	-.270**	-.030	.384**	.017	-.152*	.036	1
	Sig. (2-tailed)	.000	.696	.000	.794	.017	.572	
	Sum of Squares and Cross-products	-133.985	-488932.276	2408.220	185.932	-2779.770	573.592	4812.273
	Covariance	-.547	-2893.090	14.335	.759	-11.346	2.341	19.642
	N	246	170	169	246	246	246	246

**. Correlation is significant at the 0.01 level (2-tailed).

*. Correlation is significant at the 0.05 level (2-tailed).

Table A3. Linear regression output for the variables used in the final logistic regression model; only the collinearity statistics are relevant

Coefficients[a]

Model		Unstandardized Coefficients		Standardized Coefficients			Correlations			Collinearity Statistics	
		B	Std. Error	Beta	t	Sig.	Zero-order	Partial	Part	Tolerance	VIF
1	(Constant)	1.412	.217		6.513	.000					
	mean travel time to work (minutes)	-.010	.007	-.095	-1.532	.127	-.252	-.120	-.081	.728	1.374
	SR bus stop density (per square mile)	.007	.002	.259	3.391	.001	.597	.257	.179	.477	2.096
	total SR length (miles)	.003	.002	.102	1.855	.065	-.001	.144	.098	.921	1.086
	average annual daily SR traffic (integer)	6.796E-006	.000	.286	5.121	.000	.220	.373	.270	.891	1.122
	average SR speed limit (mph)	-.025	.004	-.416	-5.606	.000	-.564	-.403	-.296	.505	1.980
	percent of families below federal poverty level (integer)	.008	.003	.157	2.159	.032	.468	.167	.114	.528	1.894

a. Dependent Variable: SR accident rate (1 or 0)

Table A4. Final Logistic Regression Model Omnibus Tests

Omnibus Tests of Model Coefficients

		Chi-square	df	Sig.
Step 1	Step	10.087	2	.006
	Block	10.087	2	.006
	Model	138.617	6	.000

Table A5. Final Logistic Regression Model Summary

Model Summary

Step	-2 Log likelihood	Cox & Snell R Square	Nagelkerke R Square
1	91.335[a]	.560	.753

a. Estimation terminated at iteration number 7 because parameter estimates changed by less than .001.

63

Table A6. Final Logistic Regression Model Classification Table

Classification Table[a]

			Predicted		
			SR accident rate (1 or 0)		Percentage Correct
Observed			0	1	
Step 1	SR accident rate (1 or 0)	0	89	9	90.8
		1	11	60	84.5
	Overall Percentage				88.2

a. The cut value is .500

Table A7. Final Logistic Regression Model Variables

Variables in the Equation

		B	S.E.	Wald	df	Sig.	Exp(B)	95% C.I.for EXP(B)	
								Lower	Upper
Step 1[a]	Ave_AADT	.000	.000	15.358	1	.000	1.000	1.000	1.000
	Ave_SR_SPEED_LMT	-.214	.055	15.153	1	.000	.808	.725	.899
	SR_bus_stop_density_p er_square_mile	.102	.030	11.667	1	.001	1.107	1.044	1.174
	total_SR_length_miles	.035	.017	4.307	1	.038	1.035	1.002	1.070
	mean_travel_time_to_wo rk_minutes	-.131	.072	3.278	1	.070	.877	.762	1.011
	percent_families_below_f pl_integer	.128	.049	6.986	1	.008	1.137	1.034	1.250
	Constant	6.997	2.425	8.324	1	.004	1093.475		

a. Variable(s) entered on step 1: mean_travel_time_to_work_minutes, percent_families_below_fpl_integer.

Table A8. Final Linear Regression Descriptive Statistics – Census Tracts with Accidents

Statistics

		SR accident density (per square mile)	percent non white (integer)	percent workers commuting via vehicle alone (integer)	mean travel time to work (minutes)	SR bus stop density (per square mile)	total SR length (miles)	average annual daily SR traffic (integer)	average SR lane count	average SR speed limit (mph)	percent of families below federal poverty level (integer)
N	Valid	76	76	76	76	76	76	75	75	74	76
	Missing	0	0	0	0	0	0	1	1	2	0
Mean		4.6351600	39.2904	72.459	24.2066	24.780142	15.742123	48145.72969	2.539201	44.750957	16.5895
Std. Error of Mean		.5868120	3.02384	1.1595	.54699	2.5291491	1.9930826	2499.591401	.0514030	.6449849	1.47879
Median		2.9172821	29.7892	73.450	23.7500	16.116586	10.296437	48000.0000	2.539500	45.000000	13.9000
Mode		.26091[a]	3.38[a]	72.4[a]	21.60[a]	.6837[a]	1.8099	6733.3333[a]	3.0000	45.000	12.30
Std. Deviation		5.11605017	26.36119	10.1081	4.76857	22.0486106	17.2881130	21647.09652	.4451632	5.5483701	12.89178
Variance		26.174	694.913	102.174	22.739	486.141	298.879	468596787.8	.198	30.784	166.198
Skewness		1.981	.843	-.769	.842	1.146	2.302	.534	-.686	.006	1.568
Std. Error of Skewness		.276	.276	.276	.276	.276	.276	.277	.277	.279	.276
Kurtosis		4.015	-.629	.886	1.927	.679	6.689	1.222	1.170	.294	4.282
Std. Error of Kurtosis		.545	.545	.545	.545	.545	.545	.548	.548	.552	.545
Range		24.66897	94.97	52.1	27.30	91.1718	92.3875	120166.6667	2.5000	27.8571	73.80
Minimum		.26091	3.36	36.2	12.30	.6837	.0000	6733.3333	1.0000	30.0000	.00
Maximum		24.92988	98.32	88.3	39.60	91.8555	92.3875	126900.0000	3.5000	57.8571	73.80
Sum		352.27216	2986.07	5506.9	1839.70	1883.2908	1196.4013	3610929.727	190.4401	3311.5708	1260.80

a. Multiple modes exist. The smallest value is shown

64

Table A9. Final Linear Regression Variable Correlations – Census Tracts with Accidents

Correlations

		SR accident density (per square mile)	percent non white (integer)	percent workers commuting via vehicle alone (integer)	mean travel time to work (minutes)	SR bus stop density (per square mile)	total SR length (miles)	average annual daily SR traffic (integer)	average SR lane count	average SR speed limit (mph)	percent of families below federal poverty level (integer)
SR accident density (per square mile)	Pearson Correlation	1	.237	-.167	.147	.503	-.059	.052	-.295	-.288	.300
	Sig. (2-tailed)		.039	.140	.205	.000	.612	.659	.010	.013	.001
	N	76	76	76	76	76	76	75	75	74	76
percent non white (integer)	Pearson Correlation	.237	1	-.873	.217	.409	-.123	-.246	-.160	-.330	.795
	Sig. (2-tailed)	.039		.000	.060	.000	.291	.033	.170	.004	.000
	N	76	76	76	76	76	76	75	75	74	76
percent workers commuting via vehicle alone (integer)	Pearson Correlation	-.167	-.873	1	-.167	-.453	.009	.056	.003	.212	-.686
	Sig. (2-tailed)	.140	.000		.150	.000	.940	.631	.976	.070	.000
	N	76	76	76	76	76	76	75	75	74	76
mean travel time to work (minutes)	Pearson Correlation	.147	.217	-.167	1	.121	.200	.030	.038	.160	.400
	Sig. (2-tailed)	.205	.060	.150		.298	.083	.800	.749	.105	.000
	N	76	76	76	76	76	76	75	75	74	76
SR bus stop density (per square mile)	Pearson Correlation	.503	.409	-.453	.121	1	-.072	-.167	-.233	-.543	.502
	Sig. (2-tailed)	.000	.000	.000	.298		.535	.152	.044	.000	.000
	N	76	76	76	76	76	76	75	75	74	76
total SR length (miles)	Pearson Correlation	-.059	-.123	.009	.200	-.072	1	.045	-.111	.403	.044
	Sig. (2-tailed)	.612	.291	.940	.083	.535		.703	.343	.000	.708
	N	76	76	76	76	76	76	75	75	74	76
average annual daily SR traffic (integer)	Pearson Correlation	.052	-.246	.056	.030	-.167	.045	1	.492	.402	-.123
	Sig. (2-tailed)	.659	.033	.631	.800	.152	.703		.000	.000	.292
	N	75	75	75	75	75	75	75	75	74	75
average SR lane count	Pearson Correlation	-.295	-.160	.003	.038	-.233	-.111	.492	1	.178	-.130
	Sig. (2-tailed)	.010	.170	.976	.749	.044	.343	.000		.130	.236
	N	75	75	75	75	75	75	75	75	74	75
average SR speed limit (mph)	Pearson Correlation	-.288	-.330	.212	.160	-.543	.403	.402	.178	1	-.231
	Sig. (2-tailed)	.013	.004	.070	.105	.000	.000	.000	.130		.048
	N	74	74	74	74	74	74	74	74	74	74
percent of families below federal poverty level (integer)	Pearson Correlation	.300	.795	-.686	.400	.502	.044	-.123	-.130	-.231	1
	Sig. (2-tailed)	.001	.000	.000	.000	.000	.708	.292	.236	.048	
	N	76	76	76	76	76	76	75	75	74	76

*. Correlation is significant at the 0.05 level (2-tailed).

**. Correlation is significant at the 0.01 level (2-tailed).

Table A10. Linear regression output for the variables used in the final linear regression model – census tracts with accidents

Coefficients[a]

Model	Unstandardized Coefficients		Standardized Coefficients	t	Sig.	Correlations			Collinearity Statistics	
	B	Std. Error	Beta			Zero-order	Partial	Part	Tolerance	VIF
1 (Constant)	7.192	9.249		.778	.440					
percent non white (integer)	-.013	.035	-.063	-.358	.722	.269	-.045	-.034	.289	3.463
percent workers commuting via vehicle alone (integer)	.086	.072	.169	1.193	.237	-.174	.148	.113	.446	2.241
mean travel time to work (minutes)	.067	.120	.062	.557	.580	.145	.069	.053	.714	1.400
SR bus stop density (per square mile)	.073	.034	.314	2.190	.032	.513	.264	.207	.435	2.301
total SR length (miles)	-.027	.033	-.091	-.828	.411	-.071	-.103	-.078	.733	1.365
average annual daily SR traffic (integer)	7.817E-005	.000	.322	2.705	.009	.038	.320	.256	.631	1.584
average SR lane count	-4.414	1.337	-.368	-3.301	.002	-.327	-.381	-.312	.717	1.394
average SR speed limit (mph)	-.130	.134	-.140	-.970	.336	-.288	-.120	-.092	.430	2.325
percent of families below federal poverty level (integer)	.127	.078	.316	1.619	.110	.397	.198	.153	.234	4.276

a. Dependent Variable: SR accident density (per square mile)

Table A11. Global test of the linear regression for census tracts with accidents

ANOVA[a]

Model	Sum of Squares	df	Mean Square	F	Sig.
1 Regression	836.182	9	92.909	5.336	.000[b]
Residual	1114.290	64	17.411		
Total	1950.472	73			

a. Dependent Variable: SR accident rate (per square mile)

b. Predictors: (Constant), percent of families below federal poverty level (integer), total SR length (miles), average annual daily SR traffic (integer), mean travel time to work (minutes), average SR lane count, SR bus stop density (per square mile), percent workers commuting via vehicle alone (integer), average SR speed limit (mph), percent non-white (integer)

Table A12. Final Linear Regression Model Summary – Census Tracts with Accidents

Model Summary

Model	R	R Square	Adjusted R Square	Std. Error of the Estimate
1	.655^a	.429	.348	4.17262245

a. Predictors: (Constant), percent of families below federal poverty level (integer), total SR length (miles), average annual daily SR traffic (integer), mean travel time to work (minutes), average SR lane count, SR bus stop density (per square mile), percent workers commuting via vehicle alone (integer), average SR speed limit (mph), percent non-white (integer)

APPENDIX B:

ADDITIONAL FIGURES

```
Step number: 1
Observed Groups and Predicted Probabilities

      32 +                                                                        +
         I0                                                                       I
         I0                                                                       I
F        I0                                                                       I
R     24 +0                                                                       +
E        I0                                                                      1I
Q        I0                                                                      1I
U        I0                                                                      1I
E     16 +0                                                                      1+
N        I0                                                                      1I
C        I0                                                                      1I
Y        I0  0                                                                   1I
       8 +0  0                                                                   1+
         I00 01 10                                                              11I
         I00000000    0                              1      1     1          1111I
         I000000000 0 010   0 000   0  1000000 00001 00 01    1  011 100 0   0 01 111 01   1 111 1111111I
Predicted ---------+---------+---------+---------+---------+---------+---------+---------+---------+---------
Prob:    0        .1        .2        .3        .4        .5        .6        .7        .8        .9       1
Group:   000000000000000000000000000000000000000000000000000111111111111111111111111111111111111111111111111

         Predicted Probability is of Membership for 1
         The Cut Value is .50
         Symbols: 0 - 0
                  1 - 1
         Each Symbol Represents 2 Cases.
```

Figure B1: Observed Groups and Predicted Probabilities of Final Logistic Model